Thank An Animal: Oncology
By Dessi McEntee

Copyright © 2024 Ferae Projects LLC
All rights reserved.

Thank An Animal: Oncology pays tribute to the animals who sacrificed their lives in the name of developing the most powerful drugs to fight one of the most powerful diseases: cancer.

Covering well-known and not so well-known medicines, Thank An Animal: Oncology offers insights to the animals that were used to bring these anticancer drugs to market.

This book is dedicated to them.

RITUXIMAB

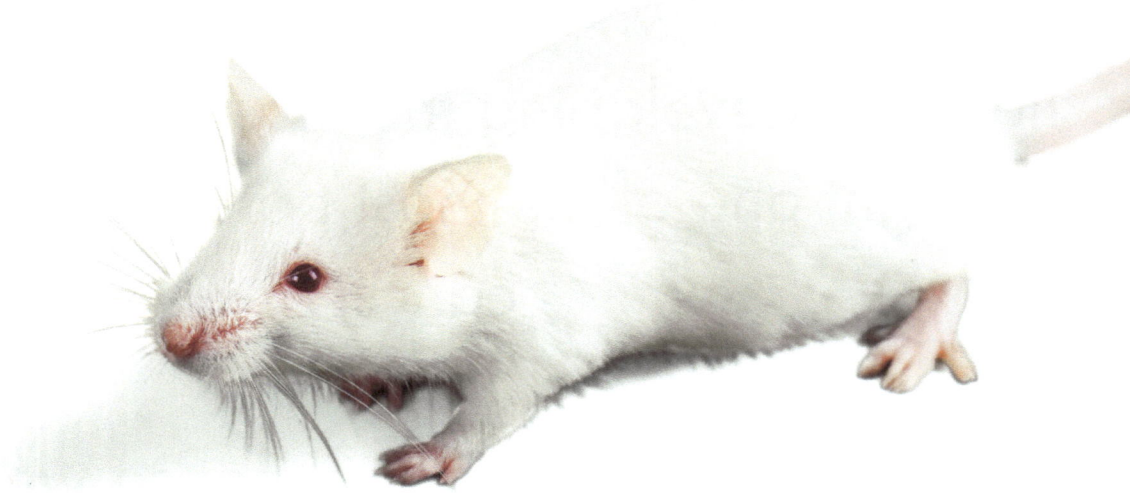

Rituximab is a monoclonal antibody used in the treatment of many cancers. It is made from a chimera of mouse and human antibodies, and is the "R" in the well-known chemotherapy treatment R-CHOP.

DOXORUBICIN

Arguably one of the most important aspects of drug development is the characterization of the inherent cardiotoxicity of a compound, or how a drug can affect the heart. Doxorubicin is a cancer drug with well-known toxic effects on the heart, and therefore it is widely used as a positive control during development within cardiotoxicity dog studies enabling accurate assessment of human cardiotoxicity risk.

MELITTIN

Melittin is not an anticancer drug, but rather a peptide found in honey bee venom that has been shown effective against several types of cancers including pancreatic cancer, one of the deadliest cancers.

#savethebees

ANASTROZOLE

Anastrozole is the most commonly used cancer drug on the market.

As an aromatase inhibitor, Anastrozole prevents the conversion of androgens into estrogen and therefore the progression of estrogen-fueled breast cancer. Cynomolgus monkeys were paramount in the development of this pivotal cancer drug.

TRASTUZUMAB

HER2 is a powerful oncogene in breast cancer and was first discovered in rats in 1984.

Fourteen years later Trastuzumab was developed as an anti-HER2 compound and is now one of the leading anticancer drugs for breast cancer.

TOZULERISTIDE

One of the more interesting medicines in oncology treatment, Tozuleristide is considered a "tumor paint" that helps paint brain tumors to make them easier to see during surgery.

To make Tozuleristide, scorpion venom is combined with a special dye.

The venom binds to brain tumors and brings the dye with it for precise imaging.

HUACHANSU

HuaChanSu is extracted from the skin of Chinese Bufo toads, and many scientific studies have shown that when combined with traditional chemotherapy it significantly increases cancer fighting capabilities.

SUNITINIB

In the earliest stages of drug development, tumor bearing mice are used to show efficacy, or effectiveness, of a new drug. Sunitinib is a kinase inhibitor that has been shown to significantly increase the power of oncology drugs when administered as a first dose over several days to these mice.

TRABECTEDIN

A more recently discovered compound, Trabectedin is extracted from the sea squirt.

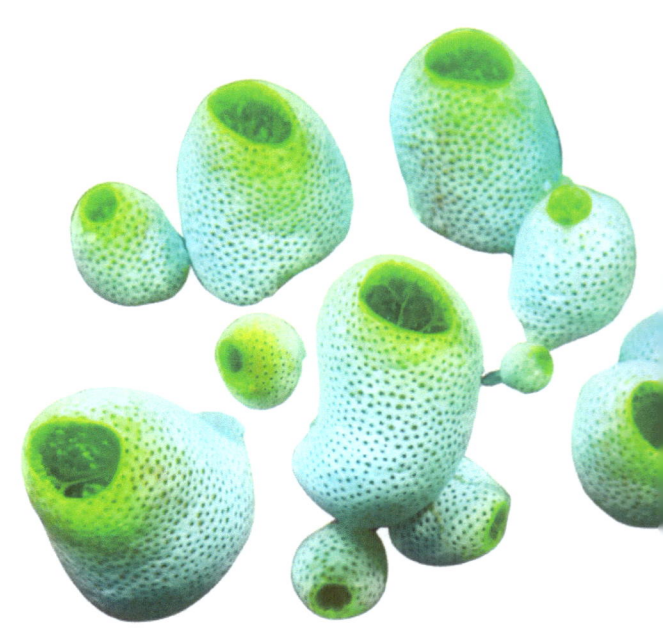

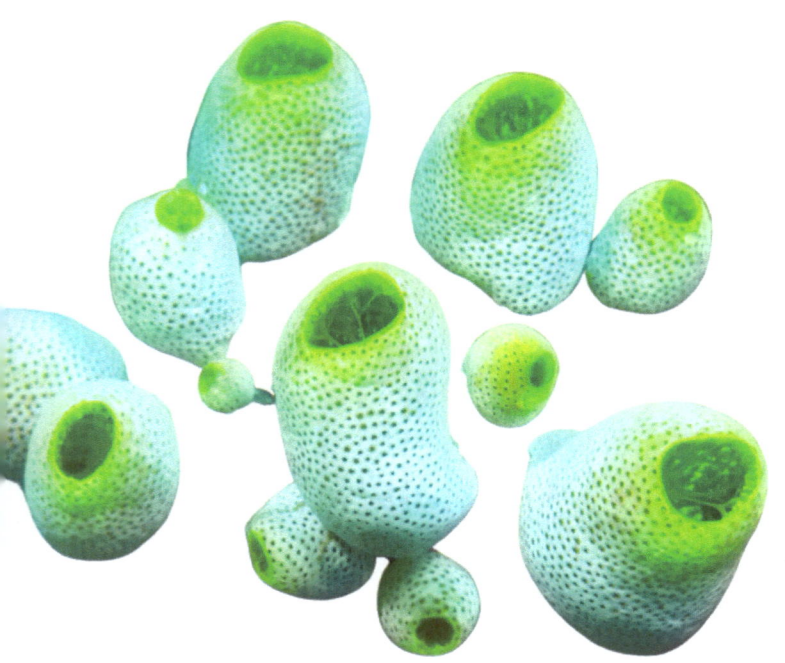

Excitedly, it is starting to show promise in battling previously drug-resistant cancers.

TCMM'S

Traditional Chinese Medical Materials (TCMM) have garnered increased attention in aiding oncology treatments.

TCMMs are natural occurring substances derived from animals such as bee venom, deer antlers, scorpion venom, sea snails and more. It is believed these materials can fight cancer through anti-proliferative effects.

RADIATION THERAPY

Many cancers are still treated with radiation therapy to kill tumors and because pig skin is very similar to human skin, pigs are essential in the development of these therapies.

PEMBROLIZUMAB

Pembrolizumab is a PD-1 inhibitor for non-small cell lung cancer. This drug was developed using a Patient Derived Xenograft model, or PDX, where tumor tissues from patients are implanted into animals like zebrafish to develop the most human-relevant medicines possible.

HU-BLT MODEL

The Hu-BLT model allows scientists to study the human immune system within the organ system of a mouse. This model, otherwise known as the human bone marrow-liver-thymus model, implants stem cells derived from human embryonic liver and thymus into the renal capsule of a mouse.

The hu-BLT model gives us a powerful engine for cancer research and using our own immune systems to fight it.

CAR T-CELL THERAPIES

CAR T-cell therapy takes a patient's own T-cells and essentially turns them into potent cancer-fighting cells. By using the patient's own T-cells, CAR T-cell therapies reduces the risk of immunogenicity which is a common effect of immunotherapy drugs. Because tumors in dogs are so common and occur spontaneously, similar to humans, dogs have been used extensively in CAR T-cell therapy development for both human and canine cancer treatments.

TAZEMETOSTAT

One of the more novel drugs to come out for cancer are epigenetic medicines. These drugs don't work by directly fighting the cancer head-on, but rather by altering the expression of genes that are over- or under- expressed in certain types of cancer.

Tazemetostat is one of these such drugs, and fights cancer by reducing gene expression of the abnormal proliferation of B-cells in follicular lymphoma (FL). We have rats and monkeys to thank for the development of this impactful drug.

TO ALL OF YOU: THANK YOU!

About the Author

Dessi McEntee is a board certified toxicologist and a published author with an MS in Pharmacology and Toxicology and a BS in Animal Science. With over a decade in the industry, Dessi has seen the good, the bad and the ugly of pharma. Today, she welcomes in the outside community towards understanding the role animals play in the scientific community.

Stay tuned for more books in the Thank An Animal series!

www.ingramcontent.com/pod-product-compliance
Lightning Source LLC
Chambersburg PA
CBHW051839210526
45473CB00005B/1945